Facing the Everyday Ethical Dilemmas of Biotechnology

# playing GOD?

Hosted by
**Charles Colson**
and
**Nigel M. de S. Cameron, Ph.D.**

---

## PARTICIPANT GUIDE
*by Tracey D. Lawrence*

Loveland, Colorado

**Playing God?: Facing the Everyday Ethical Dilemmas of Biotechnology**
Hosted by Charles Colson and Nigel M. de S. Cameron, Ph.D.
PARTICIPANT GUIDE
Copyright © 2004 The Wilberforce Forum, a division of Prison Fellowship Ministries

All rights reserved. No part of this book may be reproduced in any manner whatsoever without prior written permission from the publisher, except where noted in the text and in the case of brief quotations embodied in critical articles and reviews. For information, write Permissions, Group Publishing, Inc., Dept. PD, P.O. Box 481, Loveland, CO 80539.

Visit our Web site: www.grouppublishing.com

## *Group's* R.E.A.L. Guarantee® to you:

This Group resource incorporates our R.E.A.L. approach to ministry—one that encourages long-term retention and life transformation. It's ministry that's:
**RELATIONAL** Because learner-to-learner interaction enhances learning and builds Christian friendships.
**EXPERIENTIAL** Because what learners experience through discussion and action sticks with them up to 9 times longer than what they simply hear or read.
**APPLICABLE** Because the aim of Christian education is to equip learners to be both hearers and doers of God's Word.
**LEARNER-BASED** Because learners understand and retain more when the learning process takes into consideration how they learn best.

CREDITS

THE WILBERFORCE FORUM:
Charles W. Colson, chairman
Michael A. Snyder,
    senior vice president
Nigel M. de S. Cameron, Ph.D.,
    dean
Tracey D. Lawrence, Ph.D., M.A., project manager and writer

**BREAKPOINT** WITH *Chuck Colson*

James D. Ionkowich, D. Min, managing editor
Catherina Hurlburt, associate editor
Roberto Rivera, senior fellow, writer
Anne Morse, senior writer
Gina Dalfanzo, writer

GROUP PUBLISHING, INC.:
Beth Robinson, editor
Matt Lockhart, creative development
    editor
Joani Schultz, chief creative officer
Janis Sampson, copy editor
Jean Bruns, book designer
Joyce Douglas, print production artist
Jeff A. Storm, cover art director/
    designer
Peggy Naylor, production manager

Unless otherwise noted, Scripture taken from the HOLY BIBLE, NEW INTERNATIONAL VERSION®. Copyright © 1973, 1978, 1984 by International Bible Society. Used by permission of Zondervan Publishing House. All rights reserved.

ISBN 0-7644-2643-5
10 9 8 7 6 5 4 3 2 1     13 12 11 10 09 08 07 06 05 04
Printed in the United States of America.

# CONTENTS

Session Overview .................................................. 5

Introduction ...................................................... 7

## SESSION

**1** *Introduction to Bioethics:*
How Should Christians Deal With Making
and Taking Life? ................................................. 9

**2** *Imago Dei:*
What Does It Really Mean to Be Human? ........................... 15

**3** *Christian Ethics:*
How Do We Decide What to Think and Do? .......................... 21

**4** *Knowledge and Technology:*
Is Life Truly Better in the Information Age? .................... 27

**5** *Cloning:*
Is Cloning Good Science or Are Scientists
Playing Frankenstein? ........................................... 33

**6** *Stem Cell Research:*
Is the Chance of a Cure Worth the Price? ........................ 40

**7** *Eugenics:*
Are Super Babies Really on the Way? ............................. 47

**8** *Artificial Intelligence:*
Should We Mix Man and Machine? .................................. 53

**9** *Abortion:*
Is Early Life Less Valuable to Our World? ....................... 59

**10** *In Vitro Fertilization:*
What Dilemmas Do We Face in the Fertility Lab? .................. 66

**11** *Euthanasia:*
Who Is Less Worthy of Life? ..................................... 72

**12** *Facing the Future:*
How Can Christians Influence Bioethics? ......................... 78

This book is dedicated to the unexpected, *seemingly* least likely candidates to appear on my dedication page. I want to thank Jeff, my oldest brother, who challenged me as a young teen to have an answer for my faith, exposing the parts of my faith that were naive and unexamined—even unexamined by me. Such an admonition gave me the desire to further press the hard questions about God and to finally realize God is much bigger than my questions.

Though a bit unconventional and perhaps a borderline faux pas, I thank Madeline, my very important beagle friend, who teaches me so much about *wonder* and somehow in her charmingly doggish way, lets me know she accepts my strange humanness. Her four freckled legs help keep my two wandering feet on the ground.

To the Gamma Girls from MCC. Your wild, unabandoned prayers of hope still carry me. Thank you Kelly H., Kelly G., Cindy, and Rebecca. I haven't forgotten those God-moments with you.

With gratitude to you all and to our Creator,
*Tracey D. Lawrence*

Each session in this study is designed to help participants learn what it means to look at the world through the lens of Christianity. The sessions accomplish this through discussion, study, and relationship with the others in the group. Here's a breakdown of how each session is structured.

## GETTING STARTED *(15 minutes)*

The warm-up activity will introduce participants to the session topic. The group will participate in a fun, nonthreatening activity together and will then have a discussion about their experience. Participants will be encouraged to build relationships, to share of themselves, and to begin to think about the day's topic.

## A VIEW OF OUR WORLD *(30 minutes)*

**BREAKPOINT**

In this activity participants will listen to and discuss a BreakPoint radio broadcast narrated by Chuck Colson or Nigel Cameron. These broadcast segments are on the CD in the *Playing God?* kit.

**CULTURE**

Next, participants will watch and discuss a video segment from the video found in the *Playing God?* kit. In these video segments, Chuck Colson and Nigel Cameron will give the group more information about the day's topic.

**QUOTE.**

Then participants will read and discuss a quotation that relates to the day's topic.

## THE VIEW FROM SCRIPTURE *(30 minutes)*

At this point the group will dig into the Scriptures to see what the Bible says about the day's topic.

## WRAPPING UP *(15 minutes)*

This section includes a closing activity to challenge and encourage participants to draw conclusions from what they've learned and to put them into practice. There is also a time for group prayer.

## A VIEW OF YOUR WORLD

In this section, participants will be encouraged to do something during the week that will help them apply what they've learned to their own life situations. Participants will be asked to report on what they did and what they learned at the beginning of the next session.

# INTRODUCTION

How many times have you found yourself in a discussion about bioethics? You may initially think not many, but perhaps it's more times than you realize. Without looking too hard, we can encounter bioethics at many levels of discussion in various platforms of our culture, such as politics, movies, television, newspapers, the classroom—and even in our own lives. Do you know someone who is grappling with "pulling the plug" for an aging parent on life support? Maybe you and your spouse are exploring the option of in vitro fertilization. Or perhaps your neighbor's daughter is faced with an unwanted pregnancy and feels pressure from her family to have an abortion. Bioethics is really all about what it means to be human and why we're unique and set apart from the rest of creation. As Christians, we need to be examining how we're affected by our culture and how it's defining who we are.

This study will look at bioethics at two levels: the making of life and the taking of life. How do we bear the responsibility of creating life in a test tube, and what are the consequences of medically ending a life? In this twelve-session study, we'll cover issues such as abortion, euthanasia, cloning, and stem cell research and consider on a personal level the questions "What does this mean to me and why should I care?" We want to challenge you to move your faith from a private, internal understanding to the public square. Abraham Heschel wisely said, "The road to the sacred leads through the secular." We cannot avoid our culture and expect to influence others. The Christian can be a powerful reminder to the world that there is dignity in every life, unique to the human race alone.

As you begin this study, it might be helpful to see your small group as a think tank. Critics of Christianity may see that label as a contradiction because Christians have been portrayed as those who aren't interested in thinking about topics other than prayer and fasting. But the truth is the Christian should be more informed about all matters of life, such as bioethics, because all of life greatly matters to God. In fact, we have the inside track to truth, but we don't always acknowledge it.

This study format is discussion-driven in its design and may be a new style of learning for you. Don't think of this as just chit chat but rather a dynamic way to learn truth within community. Jesus gives us the best example of how life-changing oral tradition can be as he taught the disciples by telling stories, asking questions, and examining the Scriptures with them. The disciples were free to interject their struggles and questions, often not getting the answers they anticipated. You may feel uncomfortable at times when there is more than one answer to a question. Or you may find that you're thrown into a discussion with differing viewpoints and are unsure what you believe. Remember that the group experience is about working through the study together. It's about relationship as well as knowledge.

We pray that this group experience will challenge you to think more deeply about the issues surrounding bioethics and how your Christian worldview can transform the world around you.

> *Now for this reason also, applying all diligence, in your faith supply moral excellence, and in your moral excellence, knowledge, and in your knowledge, self-control, and in your self-control, perseverance, and in your perseverance, godliness, and in your godliness, brotherly kindness, and in your brotherly kindness, love. For if these qualities are yours and are increasing, they render you neither useless nor unfruitful in the true knowledge of our Lord Jesus Christ.*
> —2 PETER 1:5-8, NEW AMERICAN STANDARD VERSION

Honoring You in Christ Jesus,
*Tracey D. Lawrence*

# How Should Christians Deal With Making and Taking Life?

**TODAY'S TOPIC:** *Introduction to Bioethics*

## GETTING STARTED *(15 minutes)*

Take time to introduce yourself to your group, and explain why you were interested in a Bible study on bioethics and biotechnology.

Together, review the list of lesson topics in the front of your book. Then, discuss the following questions:

- Which topic do you feel you know least about, and which one do you feel you know the most about?

- Which topic is particularly relevant to you? to your church?

• What biotechnology issues have you seen in the news media recently? What makes those issues compelling news stories?

## A View of Our World *(30 minutes)*

### BreakPoint

**LISTEN TO TRACK 1 OF THE CD.** In this BreakPoint segment, which aired January 3, 2003, Chuck Colson explains where bioethics is currently leading us.

• What ethical issues are immediately raised when people have the power to create life?

• What ethical issues are raised when we are faced with the power to take life?

### Culture

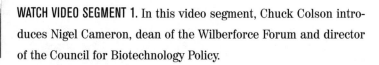

**WATCH VIDEO SEGMENT 1.** In this video segment, Chuck Colson introduces Nigel Cameron, dean of the Wilberforce Forum and director of the Council for Biotechnology Policy.

• Since *Roe v. Wade*, what new dilemmas have been raised in the field of bioethics that we did not face before?

- Do you think culture is on a "slippery slope" of increasing immorality in the area of bioethics? Why or why not?

*"Each generation exercises power over its successors: and each, in so far as it modifies the environment bequeathed to it and rebels against tradition, resists and limits the power of its predecessors...In reality, of course, if any one age really attains, by eugenics and scientific education, the power to make its descendants what it pleases, all men who live after it are the patients of that power. They are weaker, not stronger: for though we may have put wonderful machines in their hands we have pre-ordained how they are to use them."*
—C.S. Lewis

Read the quotation from C.S. Lewis, and discuss these questions:

- Do you agree or disagree with this quotation? How can too much power actually make us a weaker society?

- How does what Lewis said apply to bioethics?

# THE VIEW FROM SCRIPTURE *(30 minutes)*

Read John 3:16-17; 15:18-19; 16:33; James 4:4; and 1 John 2:15-17.

1. How does the world's system of beliefs differ from God's? How has the world's system of beliefs affected the world we live in?

2. What warnings does Scripture give us about the world?

Read Matthew 13:24-29, 36-43.

3. What good and bad seeds do you think have been sown in the field of bioethics? How are we to live side by side with people who oppose a Christian worldview?

Read Genesis 11:1-9.

4. What does this passage teach about the role of God and the proper role of people? What errors did the people of Babel make in their thoughts and in their actions?

5. How do those in biotechnology make the same mistakes as the people of Babel? What do you think will be the consequences of those mistakes?

Read Matthew 5:13-16 and Romans 12:2.

6. Why are Christians so often silent and reluctant to act in today's world?

7. As Christians, what is our role in the world of biotechnology and bioethics today?

## **WRAPPING UP** *(15 minutes)*

Spend five to ten minutes brainstorming about how the world would be different if all people lived according to a God-centered understanding of humanity. Talk about these topics:
- reproduction
- pregnant women and their unborn children
- infants and children
- the sick and suffering
- the disabled (physically or developmentally)

- the elderly
- death and dying
- in vitro fertilization

Then discuss this question:
- How can Christians renew the morality of our culture so that our world operates on a more God-centered understanding of humanity?

**CLOSE YOUR SESSION IN PRAYER.** Be sure to take the time to pray for each other. You may want to list prayer requests below so you can pray for each other during the week.

# A View of Your World

At the end of this lesson, you may be asking yourself how you could possibly have any effect on the world in the area of bioethics. Often, we dismiss the power of prayer, especially when the issue seems hopeless or out of our reach and scope of influence. This week, take time to pray for God to reveal to you personally what he wants you to do with the information you learn from this study. Meditate further on Romans 12:2, and read Ephesians 6:12. Ask God to reveal to you ways you have conformed to the world's thinking. Christians were never meant to bear the world's burdens alone. Thank God for your small group and the strength we have in the body of Christ that we can draw from daily.

## SESSION TWO

# What Does It Really Mean to Be Human?

## Today's Topic: *Imago Dei*

## Getting Started *(15 minutes)*

Take a piece of paper, and quickly draw a picture of yourself that illustrates at least one way you reflect the image of God. Then share your illustration with the rest of the group before you discuss these questions with the rest of the group:

- What does it mean to be created in the image of God? What does the dignity of humanity mean to you?

- While we bear the image of God, we are also sinners. How do you make sense of that paradox?

- How can these two truths help us frame our views on bioethics, the making and taking of life?

> **THE VIEW REVIEWED**
> *Tell the rest of the group what insights you gained from doing the "View of Your World" activity from the last lesson.*

## A View of Our World *(30 minutes)*

**LISTEN TO TRACK 2 OF THE CD.** In this BreakPoint segment, which aired February 6, 2003, Chuck Colson discusses some positive and negative examples of biotechnology.

- What other positive uses of biotechnology can you name?

- What are some areas of biotechnology that might compromise the truth that people bear the image of God?

### CULTURE

**WATCH VIDEO SEGMENT 2.** In this video segment, Chuck Colson discusses how the dignity of humanity is being challenged in our society.

- What disrespectful views of humanity do you see in popular culture?

- Do you think pop culture's message is affecting how you view yourself? How you view others? How society treats people? Explain.

### QUOTE.

**"***If God is dead, now is the time for the elevation of humanity. This possibility—even necessity—is predicated on the fact that if God is dead we must also regard as illusory any belief in an objective moral order.* **"**
—NIETZSCHE

Read the quotation from Friedrich Nietsche, and discuss these questions:

- How did Nietzsche's view of God directly affect his understanding of humanity?

- In what ways is our society elevating humanity above God? What effect does that have on our lives?

## THE VIEW FROM SCRIPTURE *(30 minutes)*

Read Psalm 8.

1. Why do you think David is astonished by God's high regard for people? Do you find these truths astonishing as well? Explain your answer.

2. How should we position ourselves in relation to God and the rest of creation?

3. How does humanity's dominion over the earth apply to bioethics and human dignity?

Read Luke 8:40-48; 19:1-10; and Philippians 2:6-8.

4. How did Jesus measure the worth of a human life? How did he interact with the unlovely?

5. How does our culture measure a person's worth? What criteria are often used that contradict Jesus' worldview?

Read James 2:1-13.

6. Why does James oppose favoritism?

7. How does favoritism play into the field of bioethics? How does James challenge those views?

8. How does James show us the importance of seeing all humans as image bearers of God?

## WRAPPING UP *(15 minutes)*

Often we're unable to embrace how much God loves us, causing us to not accept the fact we are image bearers of God. Have someone in the group read theologian Robert Frost's comments aloud.

Then have another person read Isaiah 40:11 aloud, and reflect together on how God's love gives us our worth.

Frost writes: *"I am so glad Jesus didn't suggest they group all the children together for a sort of general blessing because he was rather tired. Instead, he took time to hold each child close to his heart and to earnestly pray for them all...then they joyfully scampered off to bed"* (as quoted in *A Glimpse of Jesus* by Brennan Manning).

**PRAYER**

**CLOSE YOUR SESSION IN PRAYER.** Be sure to take the time to pray for each other. You may want to list prayer requests below so you can pray for each other during the week.

# A View of Your World

Make a point to reach out to someone in your church or community that is viewed as "less than" due to age, social status, health condition, or for some other reason. You may want to spend time visiting the residents of a nursing home, helping underprivileged children in an after-school program, or providing needed supplies for those in a substance-abuse treatment center. You could do this activity on your own, or you could make it a group project and reach out to the shut-ins, the poor, or the sick in your congregation. Afterward, note how you encountered dignity in the lives of those you visited and how it affected you.

# How Do We Decide What to Think and Do?

## Today's Topic: *Christian Ethics*

## Getting Started *(15 minutes)*

Have someone in the group read the information in the box aloud. Then discuss the questions that follow.

> During the summer of 2000, The Minnesota Daily, a University of Minnesota student newspaper, printed an advertisement soliciting egg donors who fit the following criteria:
> - 5'6" or taller
> - Caucasian
> - high ACT or SAT score
> - college student or graduate under 30 years of age
> - no genetic medical illnesses
> - extra compensation for a gifted athlete, a science or math student, or an accomplished musician
>
> The ad promised compensation of $80,000 to the right donor.

- What general ethical issues are involved in this situation?

- Do you think it would be immoral for a woman to be compensated in this way? Why or why not?

- How do Christian ethics affect our response to such an advertisement?

> **THE VIEW REVIEWED**
> *Tell the rest of the group what insights you gained from doing the "View of Your World" activity from the last lesson.*

## A View of Our World *(30 minutes)*

**LISTEN TO TRACK 3 OF THE CD.** In this BreakPoint segment, which aired August 5, 2002, Chuck Colson discusses why it's important for Christians to feel the freedom to exercise the intellect when defending morality.

- Although this broadcast was geared to college students, why is it important for all Christians to be intellectually prepared to defend what shapes our ethics?

- What fears and fallacies have you faced when discussing your faith in the public sector?

**CULTURE**

**WATCH VIDEO SEGMENT 3**, which discusses how it is impossible to find the answers to modern moral dilemmas without God.

- Do you agree with Nigel Cameron that we cannot be truly moral people without God? Why or why not?

- Are there any signs in our society that indicate godless morality doesn't work? Explain.

**QUOTE**

"*What we do with what we know is what Christian knowing is all about.*"
—Os Guinness

SESSION 3: *How Do We Decide What to Think and Do?*

Read the quotation from Os Guinness, and discuss these questions:

• What does Os Guinness mean? Do you agree? Explain.

• Do you live according to your convictions? Explain. When do you find it difficult to live according to your beliefs?

## THE VIEW FROM SCRIPTURE *(30 minutes)*

Read Isaiah 1:17 and Galatians 5:13-14.

1. What is the basis for Christian ethics?

2. Why are these ethics so important to living out the Christian faith?

Read 2 Corinthians 10:1-5 and Ephesians 6:10-18.

3. Why is what we think so important?

4. What ideas can you pull from these passages about how we are to wage the war against anti-Christian thought?

Read John 14:26; Ephesians 5:8-10; Philippians 1:19; 1 Peter 3:15; and 2 Peter 1:5-8.

5. How does knowledge equip us to live out the will of God in modern situations?

6. Many issues in bioethics deal with thinking through the conflicts that arise when worldviews collide. What can the church do to develop the knowledge and critical thinking skills needed to make a difference in how we live and how we present our ideas to the world?

7. How can we resist the temptation to keep our faith private and instead let it influence the world's standard of ethics?

## Wrapping Up *(15 minutes)*

Tell group members about a time you faced an ethical dilemma. Tell them what you did and whether or not you feel you did the right thing. Also share what role your faith played in the decision. Talk about how hard it is to do the right thing and also how hard it can be to determine what the right thing is. Encourage each other to follow your faith as you face ethical decisions every day.

## Prayer

**CLOSE YOUR SESSION IN PRAYER.** Be sure to take the time to pray for each other. You may want to list prayer requests below so you can pray for each other during the week.

## A View of Your World

Watch *The Cider House Rules*, a movie that deals with ethics in matters of life and death, supporting the pro-choice worldview. Make note of how ethics are essential in the way we shape our worldview. Then consider:

- How do ethics come into play in this movie?
- How would Christian ethics change the outcome of the story?

## SESSION FOUR

# Is Life Truly Better in the Information Age?

## TODAY'S TOPIC: *Knowledge and Technology*

## GETTING STARTED *(15 minutes)*

Think of various inventions that have changed history and affected society significantly. Refer to the list below to begin your discussion. For every advantage, think of some of the disadvantages.

- the automobile
- electricity
- television
- the computer
- the printing press
- artificial insemination

Then discuss the following questions:

- How has technology influenced the church?

- What ethical issues do Christians face with regard to technology?

> **THE VIEW REVIEWED**
>
> *Tell the rest of the group what insights you gained from doing the "View of Your World" activity from the last lesson.*

## A View of Our World *(30 minutes)*

**BreakPoint**

**LISTEN TO TRACK 4 OF THE CD.** In this BreakPoint segment, which aired March 4, 2003, Chuck Colson discusses how technology influences the way we think and live.

- How has the computer age changed what we do? Discuss positive and negative changes.

- What are some of the moral dilemmas posed by cyberculture?

**Culture**

**WATCH VIDEO SEGMENT 4.** In this video segment, Chuck Colson discusses what technology can do to the quality of life we live.

- What do you think about Scott Savage's change in lifestyle?

- Do you think our society views most technological advancement as "mythic"? Why or why not?

## Quote, Unquote

"*Science can purify religion from error and superstition. Religion can purify science from idolatry and false absolutes.*"
—POPE JOHN PAUL II

Read the quotation from Pope John Paul II, and discuss these questions:

- What does this quote mean to you in relation to technology?

- What do you think our society knows about the cost of technologies?

# THE VIEW FROM SCRIPTURE *(30 minutes)*

Read Genesis 2:15-17.

1. How did the knowledge of good and evil change Adam and Eve? When is the pursuit of knowledge bad for us?

2. What technological advancements have the potential to challenge your dependency on God?

Read 1 Timothy 6:20-21 and 2 Peter 3:17-18.

3. With so much information available to us, how do we decide what is useful and what is harmful?

Read Ecclesiastes 1:2-9, 12-18; 4:4-6; and 5:10-11.

4. What can we learn about man's nature from these passages?

5. How do technology and the inappropriate pursuit of knowledge lead to dissatisfaction with life?

6. In what ways does technology debase humanity? In what ways does technology provide opportunities to promote the dignity of humanity?

7. What ethical boundaries should we consider as we continue to pursue knowledge and advance technologically?

## WRAPPING UP *(15 minutes)*

In this age of information, we are bombarded with high volumes of knowledge and power with very few boundaries imposed on us.

Together, make a list of all the technologies you currently rely on daily. Discuss whether you feel you have a balanced, biblical worldview that you apply to these advancements.

Then talk about what you'd each like to change about your interaction with technology.

To close this session, work together to write a code of ethics for a specific discipline of technology. For example, you might write a list of "commandments" for your interaction with computers. Such a list might begin, "I will not use my computer at the expense of my family time."

**PRAYER TIME**

**CLOSE YOUR SESSION IN PRAYER.** Be sure to take the time to pray for each other. You may want to list prayer requests below so you can pray for each other during the week.

## A View of Your World

During the first half of this week, take note of each time you use a technology that takes the place of human interaction—for example, using the bank's automated teller service rather than speaking to a bank teller. Another example is sending an e-mail to a friend rather than meeting the friend in person. During the second half of the week, choose to "fast" from one or more technologies—turn off the television, leave your cell phone at home in a drawer, play no video games. Choose instead to meet with people without the aid of technology. At the end of the week, consider what advantages and disadvantages there are to living with technology. Consider whether it would be an ethical choice to give up some technological advances full time.

SESSION FIVE

# Is Cloning Good Science or Are Scientists Playing Frankenstein?

## Today's Topic: *Cloning*

## Getting Started *(15 minutes)*

Discuss this question:

What would your life be like if a clone of one of the following people existed? Be sure to talk about both the good and the bad.
- yourself
- your spouse
- your boss
- your child
- your parents

Then discuss the following questions:

- How is cloning presented in movies? in the news media?

- Why are people so fascinated with the idea of cloning?

> **THE VIEW REVIEWED**
> *Tell the rest of the group what insights you gained from doing the "View of Your World" activity from the last lesson.*

## A View of Our World *(30 minutes)*

**BreakPoint Exercise**

**LISTEN TO TRACK 5 OF THE CD.** In this BreakPoint segment, which aired January 25, 2002, Nigel Cameron discusses human cloning.

• Nigel Cameron warns that we must win the cloning battle. Why do you think he feels that if we lose this battle, the next bioethical battles will be harder?

• Reflect on some of the arguments for human cloning you have heard. How does human cloning claim to serve us?

**Culture Watch**

**WATCH VIDEO SEGMENT 5.** In this video segment, Nigel Cameron and guest scientist David Prentice further discuss the multiple ethical issues we must face if we pursue human cloning.

• What theological issues were raised in the video? On what points did you agree or disagree with what was said?

• Discuss some of the practical difficulties raised in the video. How can these practical issues be used as an avenue to talk with non-Christians about the ills of cloning?

**QUOTE, UNQUOTE**

"*It was on a dreary night of November that I beheld the accomplishment of my toils. With an anxiety that almost amounted to agony, I collected the instruments of life around me, that I might infuse a spark of being into the lifeless thing that lay at my feet...I saw the dull yellow eye of the creature open; it breathed hard, and a convulsive motion agitated its limbs...His limbs were in proportion, and I had selected his features as beautiful...his hair was lustrous black, and flowing; his teeth of a pearly whiteness; but these luxuriances only formed a more horrid contrast with his watery eyes...I had worked hard for nearly two years, for the sole purpose of infusing life into an inanimate body...but now that I had finished, the beauty of the dream vanished, and breathless horror and disgust filled my heart.*"

—VICTOR FRANKENSTEIN

**QUOTE, UNQUOTE**

"*Like Adam, I was apparently united by no link to any other being in existence; but his state was far different from mine in every other respect. He had come forth from the hands of God a perfect creature, happy and prosperous, guarded by the especial care of his Creator; he was allowed to converse with and acquire knowledge from beings of a superior nature, but I was wretched, helpless, and alone...Why did you form a monster so hideous that even you turned from me in disgust? God, in pity, made man beautiful and alluring, after his own image; but my form is a filthy type of yours, more horrid even from the very resemblance. Satan had his companions, fellow devils, to admire and encourage him, but I am solitary and abhorred.*"

—FRANKENSTEIN'S MONSTER

SESSION 5: *Is Cloning Good Science or Are Scientists Playing Frankenstein?*

Compare what Frankenstein says with what his creature says. Then discuss these questions:

- Why do you think Victor Frankenstein was immediately disgusted over his work? How do Frankenstein's burdens flow into the bioethics of cloning?

- What was the source of the monster's suffering? How does that apply to the ethics of cloning?

## THE VIEW FROM SCRIPTURE *(30 minutes)*

1. From what you've read and learned in this lesson, what are some purposes for cloning humans? Is cloning inherently wrong, or do the purposes for the clones affect whether it is ethical?

2. How does cloning show disrespect for life? How does it challenge the idea of *imago Dei*?

Read Matthew 25:40 and 1 Corinthians 6:19-20.

3. How does a Christian worldview assume responsibility for the care of all human life?

Read Genesis 2:7; Psalms 100:3-5; 127:3; and Acts 17:24-28.

4. Is it God's exclusive right to create human life? Why or why not?

5. How is cloning different ethically from normal procreation?

6. In what way is it "playing God" for humans to dabble in cloning? What consequences might occur when humans overstep their boundaries and start doing things that only God has a right to do?

7. We live in a culture that emphasizes the ability to exercise rights. How does embryonic cloning exercise our "rights" at the expense of God's plan for embryonic life?

## WRAPPING UP *(15 minutes)*

Though Mary Shelley wrote *Frankenstein* early in the nineteenth century, it's still remarkably relevant to the ethical dilemmas of today's scientific world. To close today's lesson, work together, taking into account all you've learned and discussed, and create a plot summary for a new horror story based on the ethical dilemmas presented by experiments in human cloning. Use *Frankenstein* as your model. Create a setting and characters, including a monster. Develop an ethical dilemma that will drive the plot, and also develop a moral theme that your cautionary tale will explore. Discuss what you'd like people to learn from your story.

## PRAYER TIME

**CLOSE YOUR SESSION IN PRAYER.** Be sure to take the time to pray for each other. You may want to list prayer requests below so you can pray for each other during the week.

# A View of Your World

Read the book *Brave New World*. Consider how society is affected when human beings are "manufactured." Then visit a mall, a park, or other public place where you can watch people for a half-hour or so. Consider ways in which the dignity of humanity is present in our society and where it is lacking. Think about how Christians can work to preserve the dignity of humanity both in society and in the individual.

## SESSION SIX

# Is the Chance of a Cure Worth the Price?

## Today's Topic: *Stem Cell Research*

## Getting Started *(15 minutes)*

Read and discuss the following story and questions as they occur.

- Suppose that the one thing you want above all other things is to have a child. After years and years of prayer, you find out that you're expecting a precious new baby to be born into your family, and you praise God for his graciousness.

  - When in your life have you anticipated the birth of a baby?

  For months before the birth, you and your spouse discuss names, nursery themes, preschools, and even colleges! You remodel your home to make it child-friendly. You build a playground in the back yard and interview potential babysitters. When your baby arrives, you're thrilled. But a few months after your child's birth, you notice that things aren't quite right. Worry creeps into your perfect life. You keep hoping that you're wrong, but a long series of doctor's visits proves that you're not. Your baby is diagnosed with a rare genetic disorder, and there's little hope. Chances are, your baby won't live for more than a few short years. No matter how many doctors you visit, you get the same answer—there's nothing that can be done.

- What would you do? Would you have responded differently if you found the genetic disorder through prenatal testing?

- What makes the ethics of biotechnology so difficult sometimes?

> **THE VIEW REVIEWED**
> *Tell the rest of the group what insights you gained from doing the "View of Your World" activity from the last lesson.*

## A VIEW OF OUR WORLD *(30 minutes)*

**BREAKPOINT EXERCISE**

**LISTEN TO TRACK 6 OF THE CD.** In this BreakPoint segment, which aired December 2, 2002, Chuck Colson applauds a prime time television episode in which pop culture presented both sides of the stem cell research debate.

- What ethical dilemmas were raised in the *Law & Order* episode about stem cell research?

- If you were given the opportunity to persuasively present these issues to an audience, what would you say?

**CULTURE WATCH**

**WATCH VIDEO SEGMENT 6,** which features Nigel Cameron and guest scientist David Prentice discussing the much-debated topic of stem cell research.

- What did you learn about stem cell research that you did not know before?

- How do you think stem cell research and therapy might affect you or your family in the future?

**QUOTE UNQUOTE**

*"The lives of all of us are jeopardized when life can be bought and sold, copied and replicated, altered and aborted and euthanized...We are vulnerable in a society that thinks nothing of creating a class of human beings for the purpose of lethal experimentation and exploitation."*
—JONI EARECKSON TADA

Read the quotation from Joni Eareckson Tada, and discuss these questions:

- How does embryonic research pose a threat to our society?

- What do you think should be a Christian's guiding objectives in grappling with the stem cell research debate?

## THE VIEW FROM SCRIPTURE *(30 minutes)*

Read Matthew 9:35-38; 14:13-14; Luke 7:11-17; and Revelation 21:4.

1. How did Jesus view individuals? How did Jesus view crowds?

2. Those in favor of embryonic stem cell research tend to favor a utilitarian ethic, a part of which is the idea that one person's increased pleasure can be worth another's increased pain. In what ways is a utilitarian ethic different from Jesus' ethical view?

3. How does Jesus feel about the sick? What can we learn from Jesus' healing ministry that can help us in our pursuit of physical healing today?

Read Job 31:15; Isaiah 44:2, 24; 49:5; and Jeremiah 1:5.

4. What do these Scriptures say about embryonic human life?

Read Psalm 82:3-4 and James 1:27.

5. Do you think it's fair to consider embryos among the weak and defenseless? Why or why not?

Read Matthew 7:12; 25:40; and Philippians 2:3.

6. What responsibilities do Christians have toward the sick?

7. As Christians, how can we balance the need to protect human life with the need to have compassion for the sick? What response can we offer the disabled person that doesn't compromise the dignity of human life?

# WRAPPING UP *(15 minutes)*

Together, choose one of your government representatives, and draft a brief letter to him or her addressing your concerns about stem cell research. Make sure to explain why you feel the way you do. Prayerfully send the letter.

---

**CLOSE YOUR SESSION IN PRAYER.** Be sure to take the time to pray for each other. You may want to list prayer requests below so you can pray for each other during the week.

**PRAYER TIME**

# A VIEW OF YOUR WORLD

Sometimes how we feel about controversial issues can change depending on whether we're talking about general principles or talking about an individual for whom we care deeply. This week get involved in the lives of the terminally ill or the very young. Volunteer to help for an evening at a hospice, a nursing home, a hospital, or a nursery. You may simply want to ask permission to visit with the patients and offer to pray with them. Pray that God would increase your compassion and love for those who are suffering and for those who are young and defenseless. Consider how Christlike compassion affects your attitude toward stem cell research.

SESSION SEVEN

# Are Super Babies Really on the Way?

## Today's Topic: *Eugenics*

## Getting Started *(15 minutes)*

Have someone read the following paragraph aloud:

Suppose that you are in charge of a brand new nation. You will be establishing everything about this society on biblical values. Consider what character traits you would look for as you recruit people to live in your country. These people are to populate this new nation, build a government, and establish laws. Remember, your goal is to create as close to a biblical society as possible.

After five minutes of discussion, answer these questions:

- Is it a noble goal to want to perfect society? Why or why not?

- Why would this experiment fail even with the best people?

> **THE VIEW REVIEWED**
> *Tell the rest of the group what insights you gained from doing the "View of Your World" activity from the last lesson.*

## A VIEW OF OUR WORLD *(30 minutes)*

**BREAKPOINT EXERCISE**

**LISTEN TO TRACK 7 OF THE CD.** In this BreakPoint segment, which aired March 7, 2002, Chuck Colson discusses some of the moral dilemmas of eugenics.

• Do you find yourself empathizing with or criticizing the mother's judgment? Why?

• If you knew you could prevent your child from inheriting a genetic illness, would you? Why or why not? What ethical issues would you face?

**CULTURE WATCH**

**WATCH VIDEO SEGMENT 7.** In this segment, Chuck Colson traces the history of eugenics.

- Are you surprised by some of the latest strides in eugenics? Why or why not?

- What boundaries should a Christian consider regarding the sanctity of life in the field of eugenics?

## Quote, Unquote

"*Suppose a woman planning to have two children has one normal child, then gives birth to a haemophiliac child. The burden of caring for that child may make it impossible for her to cope with a third child; but if the disabled child were to die, she would have another...When the death of a disabled infant will lead to the birth of another infant with better prospects of a happy life, the total amount of happiness will be greater if the disabled infant is killed. The loss of happy life for the first infant is outweighed by the gain of a happier life for the second. Therefore, if killing the haemophiliac infant has no adverse effect on others, it would, according to the total view, be right to kill him.*"

—Peter Singer

Read the quotation from Peter Singer, and discuss these questions:

- How does Singer's thinking oppose a Christian worldview?

- What practical and emotional issues are not mentioned in Singer's reasoning that may arise for a family in this situation?

## THE VIEW FROM SCRIPTURE *(30 minutes)*

1. Why do people want to have perfect lives and perfect children?

2. How might eugenics be quarreling with our Maker?

Read Genesis 3:1-6, 16-24; Ecclesiastes 7:29; and Romans 8:20-22.

3. How did the Fall affect people and the world we live in? In what ways are people trying to overcome the effects of the curse through eugenics?

Read about these people with limitations:

Moses—Exodus 4:10-12

Blind Man—John 9:1-11

Paul—2 Corinthians 12:7-10

4. How did these men's limitations affect their lives? How does God use our weaknesses and imperfections?

Read 2 Corinthians 4:16–5:9.

5. What can we learn about human existence from this passage?

6. Is physical perfection an inappropriate pursuit for Christians? Why or why not? In God's kingdom, what gives people value and purpose?

7. What does Paul mean when he says to "delight in weakness"? As Christians, how can we encourage each other to accept our limitations and weaknesses?

## Wrapping Up *(15 minutes)*

Find a partner (someone you don't live with), and share three unique things about yourself that your partner may not know. Then quickly share three things that make you different from each other. Share what you learned about each other with the rest of the group, and discuss the following questions:

• What do such differences reveal to us about humanity and about God?

• How will your attitude toward yourself and others change as a result of this lesson?

**PRAYER TIME**

**CLOSE YOUR SESSION IN PRAYER.** Be sure to take the time to pray for each other. You may want to list prayer requests below so you can pray for each other during the week.

## A View of Your World

Look through the newspaper or search the Web regarding the latest trends in eugenics. What do you find to be interesting? disturbing? What moral issues are raised? Have you thought much about this movement in science prior to this lesson? Bring your research to the next study, and share your observations with your group.

## SESSION EIGHT

# Should We Mix Man and Machine?

## Today's Topic: *Artificial Intelligence*

## Getting Started *(15 minutes)*

Robots are one application of artificial intelligence that you're probably familiar with. Together, name some robots from film, television, and literature, and discuss their impact on the plots.

• What were the purposes of each robot? Did they harm man or help him?

• How has the idea of the robot progressed in the science fiction world?

Take a piece of paper, and design a robot. Decide what its purpose would be and how much intelligence you would give it. When everyone's done, share your robot with the rest of the group and discuss this question:

• How much intelligence is too little for a robot, and how much intelligence is too much for a robot?

> **THE VIEW REVIEWED**
> *Tell the rest of the group what insights you gained from doing the "View of Your World" activity from the last lesson.*

## A View of Our World *(30 minutes)*

**BreakPoint Exercise**

**LISTEN TO TRACK 8 OF THE CD.** In this BreakPoint segment, which aired July 28, 2003, Chuck Colson discusses the controversies of creating Ramona, the first virtual woman.

- What makes Ramona so appealing to her fans?

- How is Ramona closing the gap between human and machine?

**Culture Watch**

**WATCH VIDEO SEGMENT 8** featuring Nigel Cameron, who raises the issues that await us in the dawn of the biotech age.

- Do you think artificial intelligence has the potential to threaten our definition of humanity? Why or why not?

- Compare the characteristics of a human to a machine. What's similar? different?

**QUOTE, UNQUOTE**

" *Suppose we scan someone's brain and reinstate the resulting 'mind file' into a suitable computing medium? Will the entity that emerges from such an operation be conscious? Asking that question is a good way to start an argument, which is exactly what we intend to do right here.* "
—RAY KURZWEIL

Read the quotation from Ray Kurzweil, and discuss these questions:

- Computer scientist Ray Kurzweil says we'll have the ability to download our brain into a computer in the future. What moral issues might we encounter with such a capability?

- To what extent should we mix man and machine?

# THE VIEW FROM SCRIPTURE *(30 minutes)*

Read Psalm 8:3-8; Luke 10:27; and 1 Corinthians 6:19-20.

1. What makes people human? Why will there always be a distinction between man and machine?

2. What are some of the dangers that could happen in the field of artificial intelligence without a biblical understanding of the whole human—body, soul, spirit, and strength?

Read Genesis 3:17-19; Exodus 1:8-14; 2:23-24; and Ecclesiastes 2:17-26.

3. One application of artificial intelligence is to use machines to do labor that humans find monotonous or dangerous. What role does physical labor play in the human experience? When is it appropriate or inappropriate to use machines for labor?

4. How can machines help people lead fuller lives without undermining people's dignity?

5. Another application of artificial intelligence is body parts to replace those that have failed due to injury or disease. Do you think there's a danger in replacing human body parts with synthetic or mechanical devices? At what point is "humanness" compromised?

6. Artificial intelligence is attempting to re-engineer people, making them superior "techno-sapiens." How does our biblical understanding of humanity enable us to respond to such efforts?

7. How might the proper use of artificial intelligence help Christians spread the good news about Christ? How might the improper use of artificial intelligence hinder those efforts?

## WRAPPING UP *(15 minutes)*

Discuss together a plan for how much intelligence machines should be given as technology continues to advance. Has your opinion changed since the beginning of this session? Come up with a set of guidelines for people who create machines, robots, computers, and other advanced technologies for whom the concept of artificial intelligence applies.

## PRAYER TIME

**CLOSE YOUR SESSION IN PRAYER.** Be sure to take the time to pray for each other. You may want to list prayer requests below so you can pray for each other during the week.

## A VIEW OF YOUR WORLD

Watch *Star Trek: First Contact*. Pay close attention to Data's comments about his struggle and desire to become human. Discuss how this relates to merging man and machine and which of Data's concerns could become ours. You might also consider reading the book *Prey* by Michael Crichton. Notice how nanotechnology is used. What is likely and unlikely to happen in our world? How does it support or oppose a Christian worldview?

SESSION NINE

# Is Early Life Less Valuable to Our World?

## Today's Topic: *Abortion*

## Getting Started *(15 minutes)*

Read the following hypothetical situation, and discuss the questions that follow.

A young teenage girl comes to you and decides to confide in you that she plans to have an abortion. She says she needs a ride to the clinic because her parents have kicked her out of the house and she's facing this alone. Which of the following responses do you feel would be more Christlike in the situation? What would you add or change to these views?

1. You would tell her that you don't agree with her decision but that you will be there to help her. You give her a ride, despite your own beliefs, because you feel she needs understanding and love first and foremost. You also offer her a place to stay as she heals from the procedure.

2. You would supply her with more education about abortion and explain to her how God views life. You offer to be there through the pregnancy and to help support her only if she chooses to rule out abortion. Further, you explain that you cannot give her a ride or help her with the abortion process, as you feel you would be enabling her to kill her child.

- What would be the consequences of choosing the first response? the second response?

- Do Christians respond appropriately to people who face problems like this? Why or why not?

> **THE VIEW REVIEWED**
> *Tell the rest of the group what insights you gained from doing the "View of Your World" activity from the last lesson.*

## A View of Our World *(30 minutes)*

**LISTEN TO TRACK 9 OF THE CD.** In this BreakPoint segment, which aired October 15, 2002, Chuck Colson gives a disturbing account on the pressure women feel to have an abortion.

- How does our culture try and "sell" us on the idea of abortion? How does this contradict a Christian worldview?

- Much pressure today falls on the mother, without much thought given toward the effects on the father of the child. How would his pain be similar to or different from the mother's pain?

**CULTURE WATCH**

**WATCH VIDEO SEGMENT 9.** In this video segment, Chuck Colson examines the strides Christians have gained in the fight against abortion.

- In what ways do you think society's attitudes toward abortion will change in the next thirty years?

- What lessons and encouragement can you glean from the example of William Wilberforce?

**QUOTE, UNQUOTE**

> "*How do we persuade a woman not to have an abortion? As always, we must persuade with love and we remind ourselves that love means to be willing to give until it hurts.*"
> —MOTHER TERESA

Read the quotation from Mother Teresa, and discuss these questions:
- How can we lovingly persuade others to choose life?

- What prolife efforts do you know about that are successfully persuading with love?

## The View From Scripture *(30 minutes)*

Read Job 3:16; 10:18-19; Psalm 139:13-18; and Jeremiah 20:16b-18.

1. In what ways is prenatal life significant?

2. If life begins before birth, what does that mean pragmatically for the Christian? What conclusions and moral boundaries can we establish for ourselves? for non-Christians?

3. How do you think our society responds to the concept of the sanctity of life—the belief that all life from conception to death is sacred? In what situations would our society agree? disagree?

4. Do you think the Christian conscience has become callous to our cultural norms in regard to the sanctity of life? Why or why not?

5. Christians may not agree completely on how to defend the unborn. Some may have convictions to take a more activist approach, while others may choose another course. How can we work together despite differing views?

Read John 13:34-35; 1 Corinthians 13:1-8a; Colossians 3:12-14; and 1 John 3:18.

6. How should we treat those who support a woman's right to choose? those who have had an abortion? those who are deciding whether to have an abortion? In practical ways, how can we love people we strongly disagree with?

Read Romans 12:9-21 and 1 Peter 2:9-13.

7. What are appropriate ways for Christians to both address this problem in society and to make their voices heard in the public forum?

## WRAPPING UP *(15 minutes)*

There have been more than 40 million abortions in the three decades since *Roe v. Wade*. Chances are you know someone who has had an abortion, whether she has revealed that fact to you or not.

Suppose a woman who has had an abortion were sitting in your group right now.

• What would you want her to know about what you think about abortion?




• What would you want her to know about what you think about her?




• What would you want her to know about God?




Discuss how putting a personal face on this issue changes your approach to it. Also discuss how you as individuals and as a church can both work to encourage society to adopt a more Christian worldview and how you can minister to those who have had abortions or who are considering abortion.

**PRAYER TIME**

**CLOSE YOUR SESSION IN PRAYER.** Be sure to take the time to pray for each other. You may want to list prayer requests below so you can pray for each other during the week.

# A View of Your World

Read up on the methods of abortion and become more familiar with the overall adverse effects of abortion in the life of women. Find out what ministries your church supports that enable women to find help and counseling for unwanted pregnancies. Prayerfully consider how God might want you to be involved in these ministries.

SESSION TEN

# What Dilemmas Do We Face in the Fertility Lab?

## Today's Topic: *In Vitro Fertilization*

### Getting Started *(15 minutes)*

Start off today's session with this brainstorming game. Go around the room, and have each person complete the sentence, "To me, children are…" Each person should complete the sentence in a different way; for example, one person might complete the sentence by saying "a blessing" while someone else might say "expensive." Continue to go around the room quickly until group members can't think of anything else to say that hasn't been said.

After the game, discuss these questions:

- Why do people want to have children?
- Why would some people rather not have children?
- How has society's attitude toward children changed through history?

---

**THE VIEW REVIEWED**

*Tell the rest of the group what insights you gained from doing the "View of Your World" activity from the last lesson.*

# A View of Our World *(30 minutes)*

**BreakPoint Exercise**

**LISTEN TO TRACK 10 OF THE CD.** In this BreakPoint segment, which aired April 23, 1996, Chuck Colson points out some of the challenges we will face with artificial reproduction.

- Does in vitro fertilization have the potential to redefine what "family" is? Explain.

- Do you think it's moral to have a child through artificial reproduction? Why or why not? If so, what moral vision should guide the use of IVF?

**Culture Watch**

**WATCH VIDEO SEGMENT 10,** which features Nigel Cameron discussing how our view of reproduction has changed in the technological age.

- Do people have a "right" to have children? Why or why not?

• Do you think that artificial reproduction affects our attitudes—as individuals and as a society—toward children? Explain.

> "*I am strongly opposed to the practice of creating fertilized eggs from 'donors' outside the immediate family (this would include the donation of sperm or eggs from a brother or sister of the husband and wife wishing to conceive). In my opinion, to engage in such activity would be to 'play God'—to create human life outside the bonds of marriage. I believe most conservative Christians would agree that this practice is morally indefensible from a biblical perspective.*"
> —JAMES DOBSON

Read the quotation from James Dobson, and discuss these questions:

• What do you think the Bible teaches about marriage and procreation?

• What do you think are the ethical boundaries in human reproduction?

# The View From Scripture *(30 minutes)*

1. Quickly glance through these Bible passages, and consider how these three women dealt with infertility and what happened to them as a result of their actions.

Sarai—Genesis 16:1-15
Rachel—Genesis 30:1-24
Hannah—1 Samuel 1:1-28

2. What did these women do right? wrong? How can these women encourage the childless today?

Read 1 Samuel 1:6-10 and Proverbs 3:5-6.

3. Does pursuing a treatment like IVF challenge God's sovereignty? Why or why not?

Read Philippians 4:6-7.

4. With the advancements in medicine and all the options available for infertile couples, do you think Christians might be less inclined to rely on prayer? Why or why not?

5. What ethical dilemmas do Christians face when deciding whether to pursue artificial conception?

6. What guiding principles can Christians use to help them make the right decision about IVF?

7. How can Christians encourage the world of biotechnology to treat life with dignity when dealing with artificial conception and reproduction?

## WRAPPING UP *(15 minutes)*

Together, talk about the role of children in your own lives and in the culture of the church. Discuss these topics:

- Are couples without children made to feel left out or incomplete in our group or our church? Explain.
- What responsibility do we have toward those we know who are facing infertility?
- How should we treat or counsel Christians pursing IVF?

**CLOSE YOUR SESSION IN PRAYER.** Be sure to take the time to pray for each other. You may want to list prayer requests below so you can pray for each other during the week.

# A View of Your World

If you know couples who have pursued in vitro fertilization, take time to learn their stories or read the story below.

After pursuing IVF for several years, Bill and Cheri had three children—one son and a set of twins. But they still had seven frozen embryos and faced the decision of what to do with them: donate them to science, discard them, give them away, or use them. They decided the latter two options were the only ethical choices. Cheri was forty years old and having more kids wasn't that appealing. Bill, however, didn't want to give his children to strangers. They opted to pursue another pregnancy. Knowing that only half of the embryos were likely to survive the thawing process, they decided to implant them all. Cheri lost all seven but felt at peace with the decision to give the embryos a chance to live. The experience was painful, but Bill and Cheri believed God provided them the strength to face the ethical challenges.

What was helpful for you in shaping your own moral ethics on this topic?

## SESSION ELEVEN

# Who Is Less Worthy of Life?

## Today's Topic: *Euthanasia*

## Getting Started *(15 minutes)*

Have you ever witnessed someone dying? If so, tell the group what you remember about the experience. Did the person suffer? Was he or she a loved one? How did the experience impact your perspective on death? Then discuss these questions:

- Why do people fear death?

- What is the "best" way to die?

> **THE VIEW REVIEWED**
>
> *Tell the rest of the group what insights you gained from doing the "View of Your World" activity from the last lesson.*

## A View of Our World *(30 minutes)*

**BreakPoint Exercise**

**LISTEN TO TRACK 11 OF THE CD.** In this BreakPoint segment, which aired January 24, 2002, Nigel Cameron exposes the dangerous philosophy surrounding euthanasia.

- What justification is usually raised to support euthanasia?

- What makes euthanasia such a complex issue?

**CULTURE WATCH**

**WATCH VIDEO SEGMENT 11** with Nigel Cameron, in which he discusses the tensions between the sanctity of life and the quality of life.

- How does the tension between quality of life and the sanctity of life make ethical decisions hard for some to make?

- How can Christians make a stand for life?

**QUOTE, UNQUOTE**

"*No image of the Supreme may be fashioned, save one: our own life as an image of His will. Man, formed in His likeness, was made to imitate His ways of mercy. He had delegated to man the power to act in His stead. We represent Him in relieving affliction, in granting joy. Striving for integrity, helping our fellow men.*"
—Abraham Heschel

Session 11: *Who Is Less Worthy of Life?*

Read the quotation from Abraham Heschel, and discuss the following questions:

• What moral boundaries should apply to the medical field as it seeks to relieve suffering?

• How can we strive for integrity when dealing with death and suffering?

## THE VIEW FROM SCRIPTURE *(30 minutes)*

Read Job 3:20-26 and 7:11-16.

1. What do people with this kind of despair need? Why are old age and dying causes for despair?

Read Leviticus 19:32; Psalms 71:9; 92:14; Proverbs 23:22; Zechariah 8:1-8; and 1 Timothy 5:1.

2. What do these verses teach about old age?

Read Job 14:1-2, 5; Psalms 90:10; 103:15-16; 116:15; and James 4:14-15.

3. What do these verses teach about the end of life? How does a biblical understanding of old age and death affect how we view euthanasia?

Read Psalms 23:4; 46:4; Romans 8:38-39; and Hebrews 2:14-15.

4. What does God promise to us as we grow old and die?

5. How should a Christian view death?

6. Is it right or wrong to seek to avoid suffering ourselves or to relieve suffering in others? Explain.

7. What is the ethical way to treat those who are terminally ill and suffering as they near death? What moral boundaries should we consider when facing end-of-life options?

## WRAPPING UP *(15 minutes)*

Think of a Christian you admire who endured great suffering for a prolonged season. This can be someone famous or someone within your circle of family and friends. Tell the group about this individual and how suffering affected his or her life as well as how it affected the lives of others. Together, talk about ways that God uses suffering to accomplish his purposes. Also talk about how you hope your own life will end and how you can glorify God by the way you handle end-of-life issues for yourself and your family.

**PRAYER TIME**

**CLOSE YOUR SESSION IN PRAYER.** Be sure to take the time to pray for each other. You may want to list prayer requests below so you can pray for each other during the week.

# A View of Your World

Follow some of the arguments that determine the criteria for *quality of life* in the medical field by doing a Web search. Are the standards objective or subjective? Do they seem consistent and fair? Then read about the life and ministry of Joni Eareckson Tada and how God is using her to serve others. Find resources, articles, and her comments on the Joni and Friends Web site. Log on to www.joniandfriends.org.

SESSION TWELVE

# How Can Christians Influence Bioethics?

## Today's Topic: *Facing the Future*

## Getting Started *(15 minutes)*

As a group, make a list of buzzwords or common terms in biotechnology that come to mind from participating in this series. You may just want to list some of the subjects mentioned in the sessions. Then, next to each term, add phrases that describe your own new perspective of the term or perhaps a new piece of knowledge you've learned about the topic. For example, you might list:

artificial intelligence—not just science fiction; can treat diseases

euthanasia—Hippocratic oath; not just for pets!

Review the list, and consider together what you've learned during this course.

Discuss these questions:

• In what ways has a greater understanding of these topics affected your view of bioethics?

• In light of this study, has your view of any of the biotechnology terms that you listed changed? If so, how?

> **THE VIEW REVIEWED**
> *Tell the rest of the group what insights you gained from doing the "View of Your World" activity from the last lesson.*

## A VIEW OF OUR WORLD *(30 minutes)*

**BREAKPOINT EXERCISE**

**LISTEN TO TRACK 12 OF THE CD.** In this BreakPoint segment, which aired January 28, 2002, Nigel Cameron calls Christians to prepare for the biotech century ahead.

• Do you find the new bioethics issues, such as cloning and stem cell research, any less complicated than older issues, such as abortion? Why or why not?

• How can Christians help to increase awareness about bioethics?

**CULTURE WATCH**

**WATCH VIDEO SEGMENT 12** with Chuck Colson discussing the challenges that Christians will continue to face in this bioethics age.

SESSION 12: *How Can Christians Influence Bioethics?*

- What do you feel is the most pressing area of bioethics?

- How could Christians' responses to biotechnology advances affect the world in which we live?

> *"Christians and others who wish to see an end to inhumanities, in compassion and love must offer alternative solutions to the problems...Merely to say to either one, 'You must not have an abortion'—without being ready to involve ourselves in the problem—is another way of being inhuman."*
> —FRANCIS SCHAEFFER

Read the quotation from Francis Schaeffer, and discuss the following questions:

- The church often has a lot to say about these problems but doesn't do as much to actively help solve them. Why is action so much harder than merely having an opinion?

- How can the church make a real difference in the world?

# THE VIEW FROM SCRIPTURE *(30 minutes)*

Read Matthew 5:44-48; 6:10; Colossians 4:2-6; 1 Timothy 2:1-4; and James 5:13-16.

1. Reflect for a moment on your prayer life. What specifically can you pray about in regard to the ethical problems in biotechnology?

2. How can we pray for those who hold different morals than we do? How can we make more conscious efforts to pray for our enemies?

3. How can we "make the most of every opportunity"?

Read Matthew 5:13-16; 2 Corinthians 6:3-10; 1 Timothy 6:11-21; and 2 Timothy 4:1-5.

4. To what extent is the church responsible for the spiritual and moral well-being of society? What is the church doing to represent Christ in society?

Read Proverbs 17:17; Ecclesiastes 4:12; and Acts 6:1-5.

5. How does God use community to impact our witness to the world? How are we better equipped to face bioethical dilemmas within community?

6. What can Christians do to encourage the biotech industry to make more ethical choices? How can we, the church, help those facing ethical issues concerning life and death? How does the Christian have an advantage over the secularist in dealing with bioethics?

7. What positive efforts have been made in bioethics that can be celebrated and supported?

## WRAPPING UP *(15 minutes)*

On a scale of one to ten, rate your overall awareness of bioethical issues before this study and after this study. Share with the others in your group which lesson was most helpful to you at this time in your life. What specific questions still remain from the material presented? What issues would you like to explore further?

Brainstorm some ways you can continue to stay in the know on the latest bioethical issues. What are some proactive steps you can take as a group to influence the biotechnology world to adopt a more Christian worldview?

---

**CLOSE YOUR SESSION IN PRAYER.** Be sure to take the time to pray for each other. You may want to list prayer requests below so you can pray for each other.

# A View of Your World

Read about Christians who have had a positive effect on the world around them. Consider how you can work to change your world. Here are some names to get you started:

- Charles Wesley (1707-1788)
- William Wilberforce (1759-1833)
- George Mueller (1805-1898)
- Corrie ten Boom (1892-1983)

# Help us provide you with quality worldview resources!

Thanks for using our worldview curriculum, *Playing God?*. We hope this product has been useful for you as you explore how to put your faith into action and as you disciple and mentor others in their faith.

In order for us to provide you with resources and products that can help you in your ministry, please take a few moments to complete this questionnaire and return it to us. To thank you for your time, we will send you one of our "Sources" publications.

Name _____

Address _____

City _____

State _____ ZIP _____

Phone _____

E-mail _____

How are you using this curriculum (Sunday school, small group, etc.)?

_____

What was the age group of the students in your *Playing God?* class (teen, college, young adult, adult)?

_____

What other worldview resources would you use?

_____

_____

_____

Permission to photocopy this page granted for local church use. Copyright © The Wilberforce Forum.
Published in *Playing God?* by Group Publishing, Inc., P.O. Box 481, Loveland, CO 80539. www.grouppublishing.com

\_\_ I would like information about BreakPoint Worldview Magazine.

\_\_ I would like information about Worldview Conferences.

\_\_ I would like additional information about The Wilberforce Forum.

\_\_ Sign me up to receive FREE daily BreakPoint Commentaries via e-mail.

What is your church affiliation?

_____

_____

Comments/Suggestions: _____

_____

_____

For further resources, visit our Web site at
**www.breakpoint.org**

Photocopy pages 85-86 and mail to:

**The Wilberforce Forum**
**1856 Old Reston Avenue**
**Reston, VA 20190-3305**

*The Wilberforce Forum is a division of Prison Fellowship Ministries. The mission of The Wilberforce Forum is to catalyze the church's development of a Christian worldview for the 21st century, and to resource and sustain Christians as they engage the culture to transform it for Jesus Christ. Founded by Charles Colson, The Wilberforce Forum produces the daily BreakPoint radio program heard by over 5 million people on more than 1,000 outlets, conducts worldview conferences, and produces books and study guides on worldview issues. To learn more about The Wilberforce Forum or BreakPoint, visit www.wilberforce.org or www.breakpoint.org.*